COUNTDOWN TO SPACE

PLUTO—
The Ninth Planet

Michael D. Cole

Series Advisor:
Gregory L. Vogt, Ed. D.
NASA Aerospace Educational Specialist

 Enslow Publishers, Inc.
40 Industrial Road   PO Box 38
Box 398   Aldershot
Berkeley Heights, NJ 07922   Hants GU12 6BP
USA   UK
http://www.enslow.com

Library of Congress Cataloging-in-Publication Data

Cole, Michael D.
    Pluto : the ninth planet / Michael D. Cole.
        p. cm. — (Countdown to space)
    Includes bibliographical references and index.
    Summary: Explores the planet Pluto, including its atmosphere and
composition, its early astronomical sightings, and its terrain.
    ISBN 0-7660-1953-5
    1. Pluto (Planet)—Juvenile literature. [1. Pluto (Planet)]
I. Title. II. Series.
QB701 .C65 2002
523.48'2—dc21
                                                        2001008313

Printed in the United States of America

10 9 8 7 6 5 4 3 2 1

To Our Readers: We have done our best to make sure all Internet Addresses in this
book were active and appropriate when we went to press. However, the author and
the publisher have no control over and assume no liability for the material available
on those Internet sites or on other Web sites they may link to. Any comments or
suggestions can be sent by e-mail to comments@enslow.com or to the address on the
back cover.

Photo Credits: Don Dixon, pp. 16–17, 36–37; Eliot Young et al., p. 4;
Enslow Publishers, Inc., p. 23; ©1998 Calvin J. Hamilton, p. 19; Heck's
Pictorial Archive of Art and Architecture, p. 15; Johns Hopkins University
Applied Physics Laboratory/Southwest Research Institute, pp. 27, 32;
Lowell Observatory Archives, pp. 9, 11; Lunar and Planetary Institute,
pp. 6, 20, 38; NASA, pp. 7, 12, 29; NASA/Space Telescope Science Institute,
p. 21; Paul Schenk, Lunar and Planetary Institute, p. 24; Alan Stern
(Southwest Research Institute), Marc Buie (Lowell Observatory), NASA,
European Space Agency, p. 31.

Cover Photo: NASA artwork by Pat Rawlings (foreground); Raghvendra
Sahai and John Trauger (JPL), the WFPC2 science team, NASA, and
AURA/STScI (background).

The cover photo is an illustration of what Pluto might look like.

CONTENTS

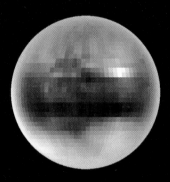

Pluto is mostly brown, shown here in this true color picture. This map of the planet was taken by recording the changes in brightness from Earth.

1

The Search for Planet X

It was January 1930. Clyde Tombaugh carefully guided the telescope to take a picture of an area of sky in the constellation Gemini. He was using one of the telescopes at Lowell Observatory in Flagstaff, Arizona.

Tombaugh had been photographing the night sky for ten months. It was difficult, detailed work, but his goal was an exciting one. Tombaugh was searching the darkness of space for a new planet.

There were eight known planets: Mercury, Venus, Earth, Mars, Jupiter, Saturn, Uranus, and Neptune. The astronomers at Lowell Observatory believed there was a ninth planet, but it would not be easy to find. For years they had been searching for what they called Planet X.

Neptune had been discovered in 1846 after observing slight irregularities in the orbital motion of Uranus.

These were the eight planets known before the discovery of Pluto: Mercury, Venus, Earth, Mars, Jupiter, Saturn, Uranus, and Neptune.

Based on the distance, size, and mass that scientists had calculated for Uranus, the planet should have followed an orbit that fit the calculations exactly. Its observed orbit, however, was slightly different than scientists had predicted. Either the scientists' calculations were wrong, or something in space was influencing the motion of Uranus.

The most likely cause was another planet. A planet's gravitational pull can be felt very far out in space from the planet. Using complex mathematical equations, astronomers calculated where this mystery planet should be. After a long and difficult search, astronomers discovered the planet Neptune. Despite its great distance from Uranus, it was Neptune that was causing Uranus to follow an orbit different from the one scientists had expected.

Not long after Neptune's discovery, Percival Lowell was studying irregularities in the orbits of comets in the outer solar system. He believed that something was affecting their orbits in the same way that Neptune affects the orbit of Uranus. He was convinced there was another planet beyond Neptune.[1]

Lowell organized his observatory and raised money to make a search for a ninth planet. In 1916, at the age of 61, Percival Lowell died. The search for Planet X was put on hold in favor of other projects at the observatory.

When Vesto Slipher became director at Lowell Observatory, he began construction of a new telescope

Percival Lowell was sure that there was another planet beyond Neptune, shown here.

and observatory dome. The telescope would have a main lens thirteen inches wide. This telescope would be used to search for Planet X.

Clyde Tombaugh Begins the Hunt

A short time later, Slipher brought in a young man from Kansas who had no professional training as an astronomer. Twenty-two-year-old Clyde Tombaugh was an avid skywatcher who had built his own telescopes. His sketches of the known planets, based on his own observations through his homemade telescopes, had impressed Slipher.

"I was taken out to the new dome and shown the new 13-inch telescope," Tombaugh later wrote. "For the first time I was told that the new telescope was to find Lowell's predicted Planet X. This sounded exciting."[2]

Tombaugh was trained to use the new telescope. He soon became an expert. He used the telescope to photograph areas of sky where calculations predicted the ninth planet might be found. Several nights later, he would photograph the same exact areas of sky again. These photographs were developed onto clear glass plates.

Tombaugh would then put the two glass plates in a device called a blink comparator. While he looked at the plates through a special microscope, the comparator would show one plate and then the other in quick succession. If the plates showed an area of nothing but

Twenty-two-year-old Clyde Tombaugh is shown with his homebuilt nine-inch reflecting telescope. He made the telescope before he started working at the Lowell Observatory in search of Planet X.

stars, his blinking view of the two plates would look exactly the same. But if anything in the photograph had moved in the sky between observations, the object would appear in a different position from one plate to the other.

After exposing more than 150 glass plates, Tombaugh had found nothing. He had also encountered many problems. The second picture of the same patch of sky could only be taken on a night when the weather was very similar to that of the night the first picture was taken. If the weather was different in the two pictures, the images of the stars would look different on the two glass plates, and the blink comparator would not work. Tombaugh also had to learn a number of tricks about guiding the telescope during the long exposure time required to make the photograph on the glass plate. Sometimes the glass plates themselves caused him problems.

"Some plates snapped in two during the one-hour exposure in the cold dome with a large BOOM, badly startling me," Tombaugh wrote.[3]

Every problem he encountered, he found a way to solve. But he had not yet found the mystery planet. An elderly astronomer who visited Lowell Observatory at the time did not have encouraging words for Tombaugh.

"Young man, I am afraid you are wasting your time," the man said. "If there were any more planets to be found, they would have been found long before this."

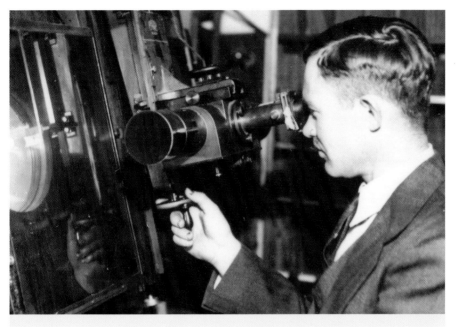

Clyde Tombaugh works the blink comparator.

But Tombaugh refused to be discouraged. "We have one of the most powerful planet-search instruments in the world," he replied. "I am going to give it all that I have."[4]

No Giving Up

Week after week, Tombaugh continued spending the night in the observatory, photographing the sky. The target area of his search moved slowly eastward through the constellations Aquarius, Pisces, Aries, and Taurus. By January 1930, after he had been at Lowell Observatory for about a year, his search had moved into a region of the constellation Gemini.

On the evening of January 23, 1930, Tombaugh photographed an area of sky around a star called Delta Gemini. He repeated the process six days later, on the night of January 29. More than two weeks passed before the plates were developed and ready to be studied. On February 18, 1930, Tombaugh sat down to study a series of plates. At 4:00 P.M., he put the two plates from January 23 and January 29 into the blink comparator and put his eye to the microscope. The pattern of stars on the photographic plate showed perfect alignment as the machine blinked the two plates in rapid succession. All the little dots of stars were in the same position on the plates from both nights—all except for one. One of the dots, smaller and dimmer than the rest, moved.

"That's it!" he said to himself.[5]

Clyde Tombaugh, a young man from Kansas with an enthusiasm for astronomy, had just discovered Planet X. But he did not let himself get carried away with excitement. Tombaugh had long ago attached a smaller

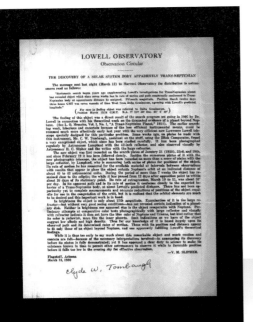

Clyde Tombaugh autographed a copy of the original Lowell Observatory announcement of Pluto's discovery.

telescope to the large thirteen-inch telescope and had taken pictures with that as well. If the moving object appeared on those plates in the same place that it appeared on the plates from the thirteen-inch telescope, it would confirm that there was nothing wrong with the equipment. It would confirm that he had found another planet.

Tombaugh placed the other plates from the smaller telescope in the blink comparator. They showed the mystery object in exactly the same place as the plates from the thirteen-inch telescope. It was "a super super moment, a moment of great elation," Tombaugh later said.[6]

He left the comparator and ran down the hall to Dr. Slipher's office. Standing at the door, he said, "I have found your Planet X."[7]

Slipher rose from his chair with a look of disbelief on his face.

"I'll show you the evidence," Tombaugh told him.[8]

They went to the comparator room and Tombaugh showed Slipher the plates. It was true. Tombaugh had discovered the ninth planet.

2

The Ninth Planet

On March 13, 1930, astronomers at Lowell Observatory announced the discovery of the new planet. The date would have been the seventy-fifth birthday of Percival Lowell.

Many different names were suggested for the new planet following its discovery. Percival Lowell's wife, Constance, had suggested calling it Percival or Lowell. Most astronomers, however, believed it needed a name from classical mythology.

Venetia Burney, an eleven-year-old girl in Oxford, England, had been studying Greek and Roman mythology, and astronomy, at her school. Because the planet was far from the Sun and very dark and cold, she thought it should be named after the Greek god of the underworld, Pluto. Her idea was sent to the Lowell Observatory. On May 1, 1930, the name Pluto became the official name for our solar system's ninth planet. The symbol for the planet is ♇. P.L. are also the initials of Percival Lowell.

The Distant Planet

The name of a god from a dark, cold underworld perfectly fits this frigid, distant planet. Pluto is indeed far from the Sun. Its orbit lies at an average distance of more than 3.6 billion miles from the Sun—39 times farther from the Sun than Earth. At that distance, Pluto takes about 248 Earth years to complete one orbit around the Sun. After being discovered in 1930, Pluto will not complete its first post-discovery orbit until August 8, 2178.

Pluto's enormous distance from Earth has made it difficult for astronomers and scientists to study. Although more than seventy years have passed since its discovery, scientists do not know nearly

Pluto was known as the Greek god of the underworld.

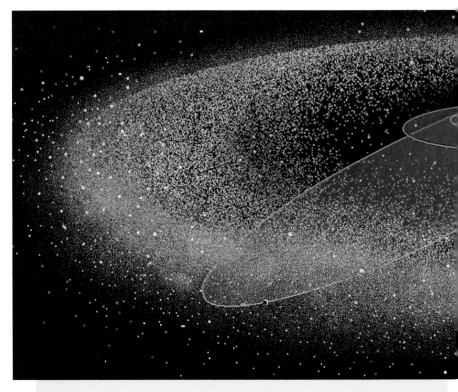

Pluto's long and tilted orbit dips above and below the orbits of all the other planets. Pluto's orbit also extends into the Kuiper Belt, pictured in blue.

as much about Pluto as they do about the other planets. At the beginning of the twenty-first century, Pluto is the only planet that has not been visited by a robotic spacecraft.

What scientists do know is that, in many ways, Pluto is different from all the other planets in the solar system.

The orbits of the other eight planets are elliptical, or oval. But Pluto's orbit is by far the most elliptical. Only

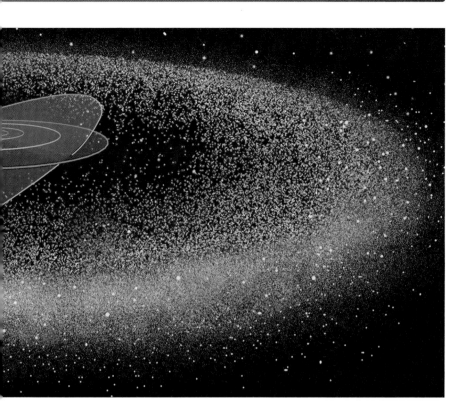

the planet Mercury comes close to having an orbit as elliptical as Pluto's. The other planets also orbit in approximately the same plane of the solar system, called the plane of the ecliptic. Pluto's orbit, however, is greatly inclined, or tilted. It is tilted 17 degrees from the plane of the ecliptic. During its orbit, Pluto dips high above and deep below the orbital plane of the other planets.

Pluto's unusually elliptical orbit sometimes carries it inside the orbital path of Neptune. But while their orbits cross, there is no danger of collision between Neptune and Pluto. In January 1979, Pluto was closer to the Sun than Neptune was. At that moment it became the eighth planet from the Sun, and Neptune became the most distant planet from the Sun. Not until February 11, 1999, did Pluto once again reach a position where it was farther from the Sun than Neptune was. Pluto will remain the ninth planet in our solar system until it passes inside the orbit of Neptune once again in the year 2227.[1]

Most of what we know about Pluto today was obtained by astronomers beginning in the 1970s. Its diameter is 1,485 miles (2,400 kilometers), compared to Earth's 7,926-mile (12,756-kilometer) diameter. The planet's small size and its distance of more than 3.7 billion miles (5.9 billion kilometers) from Earth make it difficult for scientists using telescopes on Earth to make out visible details on the planet. Earth's atmosphere also adds to the difficulty of viewing space objects from Earth's surface.

Astronomers can observe and measure slight changes in Pluto's brightness. Night after night, they have seen a pattern of changes that repeats over and over. The changes in brightness are caused by Pluto's spinning on its axis, just as Earth spins, causing night and day. Astronomers have used the information from these light changes to determine how fast Pluto spins on its axis. They found that Pluto makes one rotation every 6.4 Earth days.[2] One day on Pluto is equal to six days, nine hours, and seventeen minutes on Earth.

Pluto (left) and its moon, Charon, are shown in size relation to the United States. The planet's small size and distance from Earth make it a challenge to study.

In the 1970s, scientists began using devices called spectroscopes to analyze the light coming from Pluto. By 1976, these studies had told scientists that Pluto was made of ice and rock. The spectroscope analysis indicated that the planet's surface is covered with frozen methane, ammonia, and water ice.

Detecting these icy materials on Pluto was important. Scientists realized that Pluto's overall brightness measurement was enhanced by this very reflective icy material. Pluto's size had previously been determined by observing its brightness and relating that brightness to the planet's known distance. But scientists'

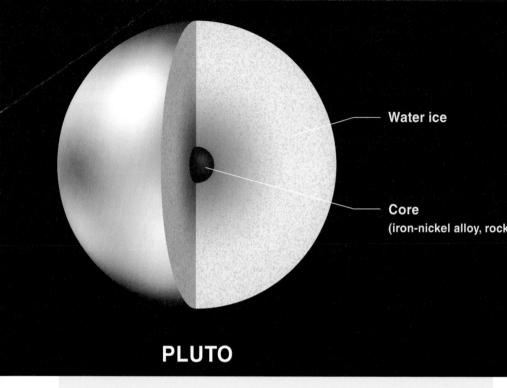

PLUTO

Scientists do not know as much about the structure of Pluto as they do about the other planets. However, machines called spectroscopes have shown that Pluto is covered with ice.

calculations had not accurately accounted for how reflective the planet was. A very large object can have low brightness if its surface is dark, and a small object can have high brightness if its surface is reflective.

The discovery of the reflective icy materials on Pluto caused scientists to change how they calculated the planet's brightness. The new calculation, with reference to the planet's known distance, resulted in a new measurement of the planet's size. It showed scientists for the first time that Pluto, a planet, is smaller than Earth's Moon.[3] It is also smaller than six other moons in our solar system.

Pluto Is Not Alone

A more startling discovery came in 1978. Astronomer James Christy was recording a series of images through his telescope at the U.S. Naval Observatory, which is also in Flagstaff, Arizona. As he studied the images, he noticed that Pluto looked lumpy. The images showed a blurry blob along the edges of the planet. Christy suspected that the images were slightly out of focus. But when he looked at the rest of the stars in the images, they were round and sharp. The blob near Pluto was not caused by a problem with the telescope. Instead, Christy suspected, something was there.

He made further observations and studies and discovered that the blob seemed to move. It appeared again in the same position every six and a half days. Later Christy observed that the total amount of light

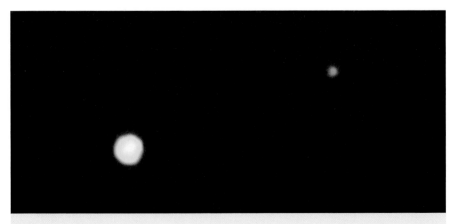

This is the clearest view yet of Pluto (left) and its moon, Charon. The Hubble Space Telescope recorded this image in 1994.

coming from Pluto periodically diminished, and that it diminished by the same amount each time.[4]

There was only one explanation: A moon was orbiting Pluto.

The drop in total light from the planet was caused when the orbiting moon passed in front of Pluto, blocking some of Pluto's reflected light. Christy had indeed discovered a moon around Pluto. It was named Charon, after the mythical boatman who ferried souls across the river Styx to be judged by the underworld god, Pluto.

Charon was later found to be approximately 736 miles (1,184 kilometers) in diameter. It is a small moon, but very large in relation to the size of its planet. It is almost exactly half the size of Pluto. Charon also appears much darker than Pluto. Scientists are not sure why. Charon makes one orbit of Pluto in the exact amount of time that Pluto takes to complete one rotation on its axis. Like Earth's Moon, the same side of Charon always faces Pluto.

Observing Charon's orbit of Pluto revealed even more about the planet.

The orbits of most moons in the solar system follow a path around their planet that is very close to the orbital plane, or the plane of the ecliptic. There is no up or down in space. But, generally speaking, a planet's north pole points "above" the plane of the ecliptic, and the south pole points "below" this plane. Most moons orbit in a region outward from their planet's equator, which is usually aligned very close to the plane of the ecliptic.

Charon, however, orbits Pluto at a steep angle above and below the plane of the ecliptic. This does not mean that Charon orbits from Pluto's north pole to its south pole. It means that Pluto's spin axis is tipped very close to the plane of the ecliptic. Charon orbits above Pluto's equator. (See illustration.) The planet Uranus is tipped on its axis in a similar fashion. As the other planets travel through their orbits, they spin like a basketball on someone's fingertip. But Uranus and Pluto move through

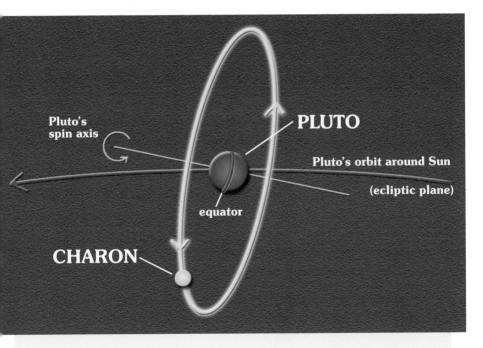

Compared to most other planets, Pluto spins on its side. Its moon, Charon, orbits Pluto above and below the ecliptic plane. Charon does, however, orbit above Pluto's equator.

space almost like a bowling ball rolling down a bowling alley. The axis of Uranus is tilted 97 degrees from the plane of the ecliptic. Pluto is tilted even farther, at 122.5 degrees.[5]

By 1985, Pluto reached a point in its orbit where something new about the planet and its moon could be observed from Earth. The angle of Charon's orbit of Pluto, and our line-of-sight view of it from Earth at that time, would allow scientists and astronomers to observe Charon passing directly in front of and directly behind Pluto. A moon passing directly in front of a planet is called a transit. A moon passing directly behind a planet is called an occultation. From these events, scientists could make detailed observations of when and how the light measurements of the two bodies changed as they moved in relation to each other. Such measurements would more accurately determine the size and mass of both Pluto and Charon.[6]

In 1985 and 1986, these transits and occultations were only partial. The first total transits and occultations occurred in 1987, and

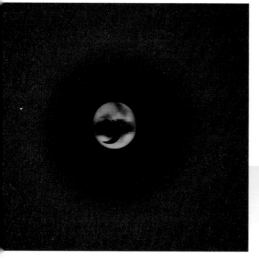

When Charon passed in front of Pluto, scientists could study the light measurements from the two objects.

scientists continued monitoring them in 1988. The studies showed scientists that Pluto is about 1,485 miles (2,400 kilometers) wide, only two-thirds the size of Earth's Moon. Charon was shown to be almost exactly half the size of Pluto, at 736 miles (1,184 kilometers) wide.

Improvements to telescopes and other astronomy-related technologies continued throughout the 1980s and 1990s. But Pluto's enormous distance from Earth and its very small size still make it impossible to observe any details on its surface. To find out more about Pluto, scientists will have to send a spacecraft there. Such a mission is planned. In the meantime, scientists have used the Hubble Space Telescope to learn more about this distant planet, and whether or not it should really be called a planet.

PLUTO[7]

Age
About 4.5 billion years

Diameter
Approximately 1,485 miles (2,400 kilometers)

Distance from the Sun
Average—3.6 billion miles (5.9 billion kilometers)
Minimum—2.7 billion miles (4.3 billion kilometers)
Maximum—4.7 billion miles (7.6 billion kilometers)

Orbital period (year)
248 Earth years

Rotation period (day)
6 days, 9 hours, 17 minutes

Surface temperature
From –390°F (–223°C) to –351°F (–213°C)

Planetary mass
1/400 the mass of Earth

Composition
A mixture of rocky and icy material

Atmospheric composition
A very thin layer of methane gas

Surface gravity
1/15 the gravity of Earth

Inclination of axis ("tilt" of the planet's axis)
122.5 degrees

Number of moons
1

New Horizons spacecraft

3

Exploring Pluto

The Hubble Space Telescope was launched into Earth orbit in 1990. From its position above Earth's atmosphere, it has been able to record images of objects in space that are far clearer and sharper than images from any telescope on Earth. In the late 1990s, the Hubble Space Telescope conducted a study of Pluto.

Hubble's images were the most detailed pictures of Pluto ever recorded. But even these images showed only the major light and dark contrasts on the planet's surface.[1]

The pictures showed for the first time that darker surfaces existed in the regions around Pluto's equator, and that bright polar ice caps were present at both of the planet's poles. Charon was a much darker object. Its

The Hubble Space Telescope took the most detailed images of Pluto ever. The telescope can take clearer pictures than telescopes on Earth's surface because it is above Earth's atmosphere.

colors indicated that it probably has a different composition and structure than Pluto.

Is Pluto Really a Planet?

The small size of Pluto and its similarity to other objects in its area of the solar system started a debate in the 1990s over whether or not Pluto should really be called a planet. In some ways, Pluto and Charon are more like asteroids. The size and composition of Pluto and Charon are more typical of asteroids. The probability that Pluto is made partly of ice makes it similar to a comet. Its orbit is certainly more like that of an asteroid or comet than that of a planet.

Some scientists think that Pluto is just one among many similar rocky icy objects in a part of the outer solar system called the Kuiper Belt. The Kuiper Belt is a region of our solar system far beyond the orbit of Neptune. It contains many icy objects, including asteroids and comets. The most massive asteroid in the Kuiper Belt, an asteroid called 2001 KX76, is as large as Charon.[2] Another asteroid, called Ceres, has a greater effect upon the orbit of comets than Pluto does. Yet Ceres is not considered a planet, while Pluto is. Why?

Defenders of Pluto remaining a planet point to the fact that Pluto has a moon. But in the mid-1990s, on its way to Jupiter, the *Galileo* spacecraft discovered a small moon in orbit around the asteroid Ida. So moons do not necessarily make the object they orbit a planet.

Comet hunter David Levy argues that Pluto cannot be called an asteroid. He explains that when Pluto was at its nearest point to Earth, the Hubble Space Telescope detected a slight atmosphere of methane hovering over the planet. Even though this may have been only the result of the Sun evaporating frozen methane off Pluto's surface, no known asteroid has an atmosphere. Levy also

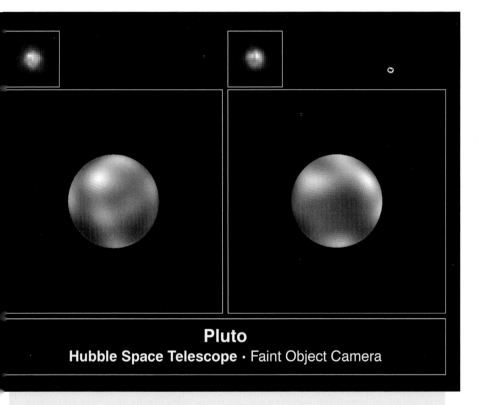

Pluto
Hubble Space Telescope · Faint Object Camera

Is Pluto really a planet? The debate continues, although the International Astronomical Union has decided it remains a planet. This image, taken from the Hubble Space Telescope in 1996, showed the never-before-seen surface of the distant planet.

The New Horizons spacecraft will visit Pluto and study its surface and its moon.

points out that Pluto is much larger than any comet astronomers have ever observed.[3]

Pluto may be larger than any comet, but its size alone is certainly not a strong point for arguing that it is a planet. Seven of our solar system's moons are larger than Pluto. Because they are in orbit around planets, they are called moons. Interestingly, the planet Mercury is smaller than one of Jupiter's moons, Ganymede, and one of Saturn's moons, Titan. Yet no one argues that Mercury is not a planet.

At least for now, Pluto seems to have survived the debate about what it is and what it should be called. It retains its classification as a planet.

Perhaps the debate will be laid to rest forever by the visit of a spacecraft to Pluto. The National Aeronautics and Space Administration (NASA) is planning a mission called New Horizons. This mission is designed to send back the first close-up pictures of the icy surface of Pluto and its darker and still very mysterious moon, Charon. Scientists plan to compare the data from these objects with what is gained from the visit to Pluto. Perhaps then they will have a better idea about whether Pluto should retain its status as a planet or be grouped together with the Kuiper Belt objects.[4]

Some scientists believe that Pluto and the Kuiper Belt objects are made of material that was left over after the formation of the other planets. Because this material has never been exposed to high temperatures from the Sun, scientists see these objects as being unchanged samples of the material that existed at the birth of the solar system. In other words, they expect that these bodies are in the same condition they were 4.5 billion years ago, when the Sun and our solar system first formed.

The New Horizons mission "represents a possible opportunity to visit the only planet not yet explored by spacecraft," said NASA's Pluto program director, Colleen Hartman. "It's really an opportunity to, in a sense, look into a deep-freeze of history which could tell us how our solar system evolved to what it is today, including the precursor ingredients of life."[5]

The New Horizons spacecraft is scheduled to launch

in either December 2004 or January 2006 and would arrive at Pluto sometime between 2014 and 2018. The arrival time depends on the launch date and the kind of launch vehicle NASA chooses to use. Once the spacecraft arrives at Pluto, its instruments will study the global geology of Pluto and Charon. Their surfaces will be mapped, and data will be collected on the planet's thin methane atmosphere. It may then move on to make similar studies on other icy, asteroid-sized bodies in the Kuiper Belt.

The New Horizons mission may provide scientists with new information that will give them a clearer picture of what the early solar system was like. The spacecraft may discover things on Pluto and Charon that change our whole understanding of those two objects. Further studies may possibly conclude that Pluto and Charon are not like other planets, Kuiper Belt objects, or comets. Perhaps scientists will have to find a whole new designation to describe this strange double world.

No matter what we call Pluto, a visit to this dark world in its remote region of the solar system would be a fascinating experience.

4

A Dark and Lonely Planet

If astronauts could ever visit Pluto, they might think that naming it after a god of the underworld was very appropriate. Pluto is so far from the Sun that the Sun's light is about 1,905 times dimmer than it appears on Earth.[1] The Sun's disk in Pluto's sky would appear only one fifth the size of the full Moon in Earth's sky.

It is unknown what the entire surface of Pluto would look like. Nearer the poles the surface would be covered with light-colored ice. Nearer the equator the surface would be darker and perhaps rocky. The entire surface would be frozen, as the temperature in many areas would be as low as –390°F (–233°C).

A wispy haze of methane gas may hang in the air. Space suits would be essential to provide warmth and

An artist shows what the frozen surface of Pluto may look like.

Pluto—The Ninth Planet

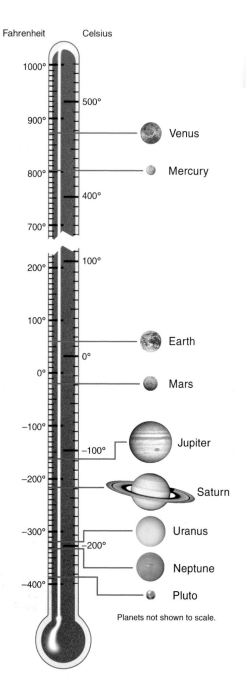

Fahrenheit
Celsius

1000°

500°
900°

Venus

800°

Mercury

400°

700°

200°
100°

100°

Earth

0°
0°

Mars

−100°

−100°
Jupiter

−200°

Saturn

−300°
Uranus

−200°

Neptune

−400°

Pluto

Planets not shown to scale.

Since Pluto is so far from the Sun, the surface temperature can be as low as −390° Fahrenheit.

oxygen to breathe. The astronauts would bounce along the surface because they would weigh only 1/15 of their weight on Earth.[2] A person weighing 100 pounds on Earth would weigh only 6.66 pounds on Pluto.

For astronauts staying on Pluto for some time, the nights and days would come very slowly. Daylight on Pluto lasts just over three Earth days, and the nights are just as long. If the astronauts were to land on Pluto where Charon is visible in the sky, the moon would never set. It would also never rise. In fact, it would always remain over the same place on Pluto. Charon's orbit is in exact relationship with Pluto's rotation on its axis. If Charon

38

were directly overhead as the astronauts stood on Pluto, it would stay there always, with its same dark face turned to them forever.[3]

Observing Pluto from Earth is something that only professional or very experienced amateur astronomers can do. Very detailed star charts and a high-quality telescope are required. The observer must consult the latest map of where to find Pluto, and then be able to distinguish its dim light from the other objects in the telescope's eyepiece. This requires very good viewing conditions and a very experienced observer at the telescope.[4]

Ask your teacher or someone at your local library if your community has an astronomy club. Members of astronomy clubs generally own their own telescopes and often give programs that allow the public to look through their telescopes at planets and other space objects. If you ask them to show you Pluto, do not be surprised if they tell you they cannot do it. Most telescopes are not strong enough to detect its dim image among the stars.

For professional and amateur astronomers alike, Pluto is a unique and difficult target for study. Until it is visited by a spacecraft, Pluto will retain its distinction as the most distant and most mysterious planet in our solar system.

CHAPTER NOTES

Chapter 1. The Search for Planet X

1. Michael E. Bakich, *The Cambridge Planetary Handbook* (Cambridge, England: Cambridge University Press, 2000), p. 299.

2. Clyde W. Tombaugh, "The Struggles to Find the Ninth Planet," *NASA Jet Propulsion Laboratory Space Science Web Page*, December 2, 2000, <http://www.jpl.nasa.gov/ice_fire//9thplant.htm> (September 29, 2001).

3. Ibid.

4. Mark Littman, *Planets Beyond: Discovering the Outer Solar System* (New York: John Wiley & Sons, 1988), p. 78.

5. Alan Stern and Jacqueline Mitton, *Pluto and Charon* (New York: John Wiley and Sons, Inc., 1998), p. 18.

6. Dennis Brindell Fradin, *The Planet Hunters: The Search for Other Worlds* (New York: Simon & Schuster), p. 80.

7. Stern and Mitton, p. 19.

8. Littman, p. 79.

Chapter 2. The Ninth Planet

1. Michael E. Bakich, *The Cambridge Planetary*

Handbook (Cambridge, England: Cambridge University Press, 2000), p. 309.

2. J. Kelly Beatty, Carolyn Collins Petersen, and Andrew Chaikin, eds., *The New Solar System* (Cambridge, Mass.: Sky Publishing Corporation, 1999), p. 290.

3. Ibid., p. 292.

4. Alan Stern and Jacqueline Mitton, *Pluto and Charon* (New York: John Wiley and Sons, Inc., 1998), pp. 52–55.

5. Bakich, p. 298.

6. Beatty, Petersen, and Chaikin, p. 292.

7. Bakich, pp. 297–298; Beatty, Petersen, and Chaikin, pp. 290–294.

Chapter 3. Exploring Pluto

1. Michael E. Bakich, *The Cambridge Planetary Handbook* (Cambridge, England: Cambridge University Press, 2000), p. 308.

2. Jeff Foust, "New Object Dethrones Ceres as Largest Minor Planet," *Spaceflight Now*, August 24, 2001, <http://spaceflightnow.com/news/n0108/24kbo/> (November 9, 2001).

3. Bakich, p. 302.

4. "NASA-Funded Study of Pluto-Kuiper Mission Completed," *Spaceflight Now*, September 29, 2001, <http://spaceflightnow.com/news/n0109/29swripluto/> (September 30, 2001).

5. "SwRI Proposal for Pluto Mission Selected by NASA," *Southwest Research Institute News*, June 13, 2001, <http://www.swri.edu/9what/releases/pluto1.htm> (November 2, 2001).

Chapter 4. A Dark and Lonely Planet

1. Michael E. Bakich, *The Cambridge Planetary Handbook* (Cambridge, England: Cambridge University Press, 2000), p. 306.

2. Paul McGehee, *The Pluto Home Page*, n.d. <http://dosxx.colorado.edu/plutohome.html> (November 4, 2001).

3. J. Kelly Beatty, Carolyn Collins Petersen, and Andrew Chaikin, eds., *The New Solar System* (Cambridge, Mass.: Sky Publishing Corporation, 1999), pp. 292–293.

4. Rick Shaffer, *Your Guide to the Sky* (Los Angeles: Lowell House, 1999), p. 90.

GLOSSARY

asteroid—Any rocky object in space that measures between a few hundred feet and several hundred miles in diameter.

blink comparator—A machine used to compare two photographic plates. By looking through a microscope at one plate and then the other, "blinking" back and forth between them in quicker and quicker succession, astronomers can see whether any of the objects on the two plates have moved from the time the first photograph was taken to the time the second was taken.

comet—A body of rock and ice in space that circles the Sun in a long, elliptical orbit. As it draws nearer the Sun, the ice is melted away, forming a visible tail of vapor.

Kuiper Belt—A disk-shaped region beyond the orbit of Neptune that contains many icy objects, including comets.

occultation—The passage of one object behind another, briefly blocking an observer's view of the object behind.

plane of the ecliptic—The plane, or flat area, in which eight of the nine planets orbit the Sun. Pluto's orbit is tilted 17 degrees from the plane of the ecliptic.

precursor—Something that comes before another.

transit—The passage of one object in front of another, in which an observer's view of the object behind is either partially or entirely blocked.

FURTHER READING

Bredeson, Carmen. *Pluto*. Danbury, Conn.: Franklin Watts, 2001.

Brimmer, Larry Dean. *Pluto*. New York: Children's Press, 1999.

Kerrod, Robin. *Uranus, Neptune, and Pluto*. Minneapolis, Minn.: Lerner Publications Company, 2000.

Vogt, Gregory. *Pluto and the Search for New Planets*. Chatham, N.J.: Raintree Steck-Vaughn Publishers, 2000.

INTERNET ADDRESSES

Arnett, Bill. "Pluto." *The Nine Planets: A Multimedia Tour of the Solar System*. April 25, 2001. <http://www.seds.org/billa/tnp5/pluto.html>.

McGehee, Paul. *Pluto Home Page*. n.d. <http://dosxx.colorado.edu/plutohome.html>.

Tombaugh, Clyde. "The Struggles to Find the Ninth Planet." *NASA Jet Propulsion Laboratory Space Science Web Page*. December 5, 2000. <http://www.jpl.nasa.gov/ice_fire//9thplant.htm>.

INDEX